FOREST CREATURES

SKUNKS IN THE FOREST

JENNIFER LOMBARDO

Please visit our website, www.garethstevens.com. For a free color catalog of all our high-quality books, call toll free 1-800-542-2595 or fax 1-877-542-2596.

Library of Congress Cataloging-in-Publication Data

Names: Lombardo, Jennifer, author.
Title: Skunks in the forest / by Jennifer Lombardo.
Description: New York, NY : Gareth Stevens, [2023] | Series: Forest creatures | Includes bibliographical references and index.
Identifiers: LCCN 2022007496 | ISBN 9781538279250 (library binding) | ISBN 9781538279236 (paperback) | ISBN 9781538279243 (set) | ISBN 9781538279267 (ebook)
Subjects: LCSH: Skunks–Juvenile literature.
Classification: LCC QL737.C248 L66 2023 | DDC 599.76/8–dc23/eng/20220303
LC record available at https://lccn.loc.gov/2022007496

Published in 2023 by
Gareth Stevens Publishing
29 East 21st Street
New York, NY 10010

Editor: Jennifer Lombardo
Designer: Rachel Rising

Portions of this work were originally authored by Jackie Heckt and published as *Skunks*. All new material in this edition was authored by Jennifer Lombardo.

Photo credits: cover Geoffrey Kuchera/Shutterstock.com; cover, p. 1 furtseff/Shutterstock.com; cover, pp. 1, 3–6, 8, 10, 12, 14, 16, 18, 20, 22–24 lavendertime/Shutterstock.com; cover, pp. 1, 3, 4, 6, 8, 10, 12, 16, 22–24 Zvezdesign/Shutterstock.com; p. 5 Eric Isselee/Shutterstock.com; p. 7 Chriscoatesphotography/Shutterstock.com; p. 7 Holly Kuchera/Shutterstock.com; pp. 9, 13 critterbiz/Shutterstock.com; p. 11 Danita Delimont/Shutterstock.com; p. 15 Design Pics Inc/Alamy Stock Photo; p. 17 Agnieszka Bacal/Shutterstock.com; p. 19 Martin Prochazkacz/Shutterstock.com; p. 21 Images Azzaro/Shutterstock.com.

Printed in the United States of America

CPSIA compliance information: Batch #CSGS23: For further information, contact Gareth Stevens, New York, New York, at 1-800-542-2595.

CONTENTS

Words in the glossary appear in **bold** type the first time they are used in the text.

NOT ALWAYS SMELLY

Everybody says skunks are stinky. This is because they have **glands** under their tail that can spray a smelly liquid. They do this only if they have no other choice. If they don't spray, they don't smell worse than any other wild animal!

There are more than 10 species, or kinds, of skunks throughout the world. Most are black and white, and almost all of them have some kind of stripe. However, only one kind is named the striped skunk.

FIND OUT MORE!

Stink badgers look like badgers, but they're actually skunks! They're the only kind of skunk that isn't **native** to North America. They live in Indonesia, Malaysia, and the Philippines.

SKUNKS AND WHERE THEY LIVE

SPECIES	HABITAT
hooded skunk	southwestern U.S., Mexico, Central America
eastern spotted skunk	eastern U.S.
American hog-nosed skunk	southern U.S. to Nicaragua
Humboldt's hog-nosed skunk	Patagonia (Argentina and Chile)
striped skunk	Canada to Mexico
Molina's hog-nosed skunk	Chile to Brazil
pygmy spotted skunk	Mexico
striped hog-nosed skunk	Mexico to Peru
stink badger	southeast Asia

STRIPED SKUNK

This table shows the habitat, or natural home, of some different skunk species.

DIFFERENT KINDS OF SKUNKS

Striped skunks are common in North America. They're found in the warmer areas of Canada, across the United States, and as far south as Mexico. Hooded skunks, which live in the southwestern United States, Mexico, and Central America, are much less common.

Spotted skunks live as far south as Costa Rica. Their name is confusing because they aren't actually spotted! Instead, the pattern of their stripes is broken, like a dotted line. The pygmy spotted skunk can fit in the palm of a person's hand.

FIND OUT MORE!

The striped skunk is the largest skunk species. It can weigh up to 14 pounds (6 kg)—about the size of a large house cat.

Striped skunks and hooded skunks look very similar, but they're different species.

SKUNK HABITATS

Skunks can live in many different habitats. Forests are especially good for them because they can use the trees and other plants as cover or shelter. Skunks can survive in very cold weather and very hot weather, which allows them to live in many different areas.

Skunks also sometimes live in grasslands and deserts. They often come into cities and **suburbs** because they smell people's food. They're commonly seen in parks and big backyards.

FIND OUT MORE!

Skunks are most attracted to garbage and pet food. Keeping your garbage in a can with a tight-fitting lid can help keep them away.

A hollow tree makes a great place for a skunk to curl up and rest.

SKUNKS AT NIGHT

Skunks are nocturnal, which means they're most active at night. Scientists say this might be why they **evolved** to spray predators. At night, they don't have to worry about hawks or most other birds. They're more likely to meet other ground animals.

Sometimes skunks come out during the day to look for food, especially if they haven't found any at night in a while. If you see one, let it be! Skunks generally aren't **aggressive**, so they won't come after you.

FIND OUT MORE!

If you see a skunk in the daytime and it's doing something animals don't normally do, it might have **rabies**. Tell an adult if this happens.

Skunks come out at night to find food and water.

SKUNK HOMES AND FAMILIES

Skunks dig holes in the ground called dens to live in. Skunks are mostly **solitary** animals, but when the weather gets colder, some skunks come together to live in one den. They keep each other warm through the winter.

A baby skunk is called a kit. A mother skunk gives birth to a litter of up to 10 kits each year. The kits are blind for about three weeks, and they depend on their mother to feed them and keep them safe.

FIND OUT MORE!

Kits are born with the ability to spray, but they can't aim their spray until they're about four months old.

Skunks are born in the spring and stay with their mother until the fall.

FINDING FOOD

Skunks are omnivores. That means they eat both plants and animals. They'll eat fruits, eggs, nuts, and grasses. They aren't picky, though—they'll gladly eat trash and **carrion**!

Skunks don't have the teeth and claws they need to kill large animals, so they eat only small things that don't put up a fight. These include snakes, rabbits, and frogs. However, their favorite foods are bugs. They use their claws to dig up beetles, crickets, worms, grasshoppers, and many others.

FIND OUT MORE!

Skunks love to eat honeybees! They trick the bees into coming out of their hive by scratching the outside of it. The skunk's thick fur keeps it safe from bee stings.

Skunks won't say no to your garbage!

LEARN THE SIGNS

A skunk's liquid spray is called musk. Skunks can spray their target from up to 10 feet (3 m) away, and the smell can travel up to 1 mile (1.6 km). They have great aim and usually shoot for the target's eyes because it hurts.

Skunks attack only if they feel like they have no other way to get rid of a predator. This is why they give a warning first. If you see a skunk stomping its feet, shaking its tail, and turning its back to you, back away slowly.

FIND OUT MORE!

Skunks warn predators first because it takes a while for their glands to make more musk after they spray. During this time, they have no **defense** if another predator attacks.

The spotted skunk's warning includes a little handstand dance!

A GOOD DEFENSE

Because of their defense, few predators attack skunks. Since they aren't killed by other animals very often, there are a lot of skunks around.

Great horned owls are one of the few natural predators of skunks. They can attack from above the skunk, which makes it harder to spray them. They also don't have a great sense of smell. Foxes, coyotes, and bobcats may try to hunt a skunk, but they will probably end up smelling for days.

FIND OUT MORE!

The best way to get skunk smell out of clothing or fur is with vinegar, baking soda, or dish soap. These things get rid of the oils in the musk that cause the smell.

Great horned owls hunt at night, when skunks are most active.

SKUNKS AND PEOPLE

Because of skunks' musk, many people don't really like skunks. However, skunks are very helpful to us, especially to gardeners and farmers. They eat a lot of bugs that ruin plants.

Some people take skunks' glands out and sell them as pets. Skunk owners say they're friendly, playful, and smart, just like cats and dogs. However, they're also likely to chew up or dig into furniture and floors. At least 33 states have laws against owning a pet skunk.

FIND OUT MORE!

Humans are one of the skunk's biggest predators. Instead of leaving them alone, some people put out traps or poison in their gardens.

Skunks can be friendly pets, but they're not for everyone.

GLOSSARY

aggressive: Showing a readiness to attack.

carrion: A dead, rotting animal.

defense: A way of guarding against an enemy.

evolve: To grow and change over time.

gland: A body part that produces something that helps with a bodily function.

native: Living or growing naturally in a particular place.

rabies: A deadly disease that affects the central nervous system. It is carried in the spit of some animals.

solitary: Tending to live or spend time alone.

suburbs: Smaller communities around a large city.

FOR MORE INFORMATION

BOOKS

Corrigan, Sophie. *The Not Bad Animals.* Minneapolis, MN: Frances Lincoln Children's Books, 2020.

McDonald, Amy. *Skunks.* Minneapolis, MN: Bellwether Media, 2022.

WEBSITES

Fun Skunk Facts for Kids
easyscienceforkids.com/all-about-skunks/
Check out even more about this awesome animal!

National Geographic Kids: Skunk
kids.nationalgeographic.com/animals/mammals/facts/skunk
See photos and read facts about skunks.

INDEX